MANUSCRIPT & INSCRIPTION LETTERS

FOR SCHOOLS AND CLASSES AND FOR THE USE OF CRAFTSMEN

By EDWARD JOHNSTON, with 5 Plates by A. E. R. GILL.

These 16 Plates are intended as a working supplement to my Handbook, "Writing and Illuminating, and Lettering" (to which frequent references are made under the letters "W. & L."), but they form a complete scheme in themselves—based on the class-sheets and class-notes given to my students during the last ten years—that I think may be of value to craftsmen and designers as well as to instructors and students.

Considerable changes have been made in some of the sheets, and some slight differences from the diagrams in the Handbook will be observed : these have all tended, I think, to come nearer to the *essential* forms. The most important simplification of form is in the Versals of Plate 5.

This scheme is indicated by the

CONTENTS.

Plates 10, 13, 14, 15, are collotypes from the actual works. The other plates are photo-lithographs : Plates 12 and 16 from drawn and written originals, Plate 11 from type, and Plates 1 to 9 from manuscript copies and examples (Plates 4, 6, 7 have been reduced ¼ linear).

Early MS. cannot be satisfactorily reproduced by a line process, nor can it be drawn. The early MSS. here given are, therefore, freely written copies rather than facsimiles, and detailed notes and references are added to increase their accuracy. They are all exactly as written (with the exception of a very few repairs), and most are absolutely untouched, so that they exhibit certain natural breaks and roughnesses, which, though a more skilful pen might avoid them, are of value in betraying to the student not only the forms, but the actual manner of their construction.

Note on the Pen : The importance of the part that the Pen has played in the development of letters cannot be over-estimated ; and I believe that it is beginning to be recognised that the best way to study letters, or even to "design" them, is to practise oneself in the use of a broad-nibbed pen, which will, after a little acquaintance with good models, *practically make the letters for us.*

For the most accurate work a sharply cut quill—or large reed—(W. & L., Chap. II.) is essential, but some valuable practice may be obtained with one of the many "round hand" steel nibs that are now sold.

A THEORY OF CALLIGRAPHY.

Alphabets wrested from their original places in MSS. and Inscriptions are in danger of becoming mere "copies," or crystallisations, that may breed literal copies and inscriptions without spirit. As an "antidote," therefore, to this collection of Alphabets, I have tried to suggest their fine beginnings, and right, or at least, sound, directions in which they may be followed out. The point of view of the early calligrapher was most direct : in the first place his Manuscript was to be read, then, to be played with or glorified The later men probably thought more consciously of "beautifying" (which is the beginning of danger), and in the last stage "Illuminators" descended to every kind of artifice. It is possible even now to go back to the child's—something like the early calligrapher's—point of view, and this is the only healthy one for any fine beginning : to this nothing can be added ; all Rules must give way to Truth and Freedom.

September, 1909.

E. JOHNSTON.

I DENY NOT, BUT THAT it is of greatest concernment in the Church & Common-wealth, to have a vigilant eye how Bookes demeane themselves as well as men; & thereafter to confine, im-prison, & do sharpest jus-tice on them as malefactors: For Books are not absolute-ly dead things, but doe con-tain a potencie of life in them to be as active as that soule was whose progeny they are; nay they do preserve as in a violl the purest effi-cacie & extraction of that living intellect that bred them. I know they are as lively, & as vigorously produc-tive, as those fabulous Dra-gons teeth; and being sown up and down, may chance to spring up armed men. And yet on the other hand unlesse warinesse be us'd, as good almost kill a Man as kill a good Book:

PLATE I.—PLAN For a MANUSCRIPT BOOK. The Book is the most practical base for the student of Manuscript Letters and, through them, of Letters generally. (Ref. W. & L., Chap. VI.)

The traditions of the modern Printed book are based on the traditions of the early Manuscript book. The Printer in his 450 years of work has given permanence to the "roman" and "italic" types and adopted the Title-page as a fixture, but he has added nothing. And though printing is, and the printed book may be, a good thing, the tools for its practice are not, like those of the scribe, at every one's hand.

PLANNING THE BOOK. The page, the margins, & the writing should as much as possible *settle themselves.*

1. Size of Page : Matter, Form, Use, Paper &c. ought to suggest this. *Paper sheets, of the common sizes & proportions (abt. 7 : 9), may determine the size & shape of page by folding for folio, 4to., 8vo., &c. Similarly, the economic cutting up of a skin may help to fix the Vellum book.*

2. Size of Margin, or proportion of Text to Page : in ordinary printed books Text=abt. ½ area of Page : in MS. Books Text=abt. ¾ area of Page (as in above Example).
   The Traditional Proportions of MARGINS to each other are, *Inner* 1½, *Top* 2, *Side* 3, *Foot* 4 +.
   Note : in a page of the common (*folio* or 8vo.) proportions of 7 : 4½ (as above Example)
   make { Height of Text column = Width of Page } { The area of Text will then = ¾ area of page and this will allow the right
   { Width of Text column = ⅔ Height of Page } { proportions of margin exactly 1½ : 2 : 3 : 4½.

3. Size of Writing : Governed generally by the *number of words to the line* : in a poem this is fixed, in prose between 4 & 8 is a convenient number (ordinary printed book = 8 or 9 : MS. Example above 5).

MAKING THE BOOK. Every process should be as swift and workmanlike as possible.

Preparing Pages : The Top edge of each *book-sheet* is cut straight and the *fold* is at right angles. *All* exact measurements are made from the Top, *downwards*, and from the *fold, outwards.*

Ruling : Lines are ruled with an indenting point, or a fine hard lead pencil. The vertical marginal lines run from head to foot of the page, the first and last *horizontals* may run across (as in Example), or stop at the vertical lines. All these lines or some of them may be doubled.

Writing : Write as straightforwardly as possible, leaving spaces for the large Initials (v. Plate 5) 1, 2, 3, or more line-spaces high according to their importance. They are generally put in *afterwards* in colour together with *rubrics &c. The intention of the original Text must be the first consideration.*

Binding : The written *book-sheets* are gathered into *quires* or *sections*, generally of 4, and sewn on to tapes or strings.

ABCDEFGHILMN

special positions
Normal pen position:

OPQRSTVXY AJKUWZ

Suggested modern A & JKUWZ forms

"SQUARE CAPITALS" freely copied with a pen from a photograph of a 4th. or 5th. Century MS. (Vergil) – abt. 2 ex. the height of originals

ACCEDENS AVTEM TRIBVNVS, DIXIT
ILLI: DIC MIHI SI TV ROMANVS ES?
AT ILLE DIXIT: ETIAM· ET RESPONDIT
TRIBVNVS: EGO MVLTA SVMMA CI-
VILITATEM HANC CONSECVTVS SVM·

Example of (modern) Writing. Note: the words are 'packed' and separate now: in the early MSS. there was no such division. (Actus Apostolorum xxii.27)

ABCDEFGHILMNOPQ
RSTVXY·AHJKUWZ

Normal pen position for 'thicks'
Suggested modern A, H, & JKUWZ forms } to match.

"RUSTIC CAPITALS" freely copied with a pen from a photograph of a 3rd. or 4th. Century MS. (Vergil) – abt. 2 ex. the height of originals.

PLATE 2.—"SQUARE" & "RUSTIC" CAPITALS the two great Book-hands of the 3rd to the 5th century A.D.
CONSTRUCTION: The Square capitals followed the inscriptional forms (cf. Pl. 12): the Rustic variety was more easily written.
Both are written with a "slanted-pen," i.e. a pen so held or cut that the thin stroke is oblique ∕ : cf. "straight pen" (Pl. 4),
the Square with a slight slant ∕, the Rustic with a great slant ∕. In either case the position is generally uniform, but
the slant is increased for all thin stems and the nib moves on to one of its "points" in making some of the terminals.
USE : The Square MS. makes a beginning book-hand ; the Rustic an occasional hand & a basis for ornamental forms (W.& L., p. 297.)

ABCDEFGhIJ
klMNOPQR
STUVXYZQ
ω
GKJ FLP
ROMAN variants without tails; see also D, E, H, utr'd in the writing below.

while the earth remaineth,
seedtime and harvest,
and cold and heat,
and summer and winter,
and day and night,
shall not cease.

PLATE 3.—" UNCIAL " LETTERS : a modern straight-pen (note 2, Pl. 4) form of the later 7th-8th century Uncials : the earlier forms were slanted pen (cf. 2, Pl. 7). Uncial writing was the chief Book-hand from the 5th to the 8th century. The letters make a fine large Book-hand used alone, or they may be used together with the Half-Uncial hand in Pl. 4 as Capitals (to be written with the same pen). They also offer a good basis for ornamental forms. (W. & L., p. 300.)

abcdefghijklmno

pqrstuvwxyz &

Varieties 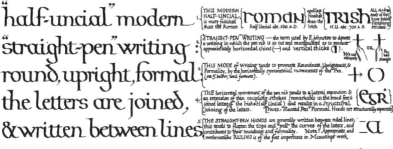 ðfgjgqr double letters ff fi fl

"half-uncial" modern_

"straight-pen" writing

round, upright, formal.

the letters are joined,

& written between lines

THIS MODERN HALF-UNCIAL is more finished than the Roman. roman and less finished than the Irish. irish Half-Uncial abt. 500 A.D. Irish H.U. abt. 700 A.D. ALL three ends of this have hidden finishing strokes.

STRAIGHT-PEN WRITING — the term used by E. Johnston to denote a writing in which the pen nib is so cut and manipulated as to produce approximately horizontal thins (—) and vertical thicks (¶). Nib cut obliquely OR. Nib held straight

THIS MODE of Writing tends to promote Roundness, Uprightness, & Formality, by the horizontally symmetrical movements of the Pen. (see § below, and footnote).

THE horizontal movement of the pen rib tends to a lateral expansion & an extension of thin coupling-strokes (remarkable in the broad feet & joined letters of the Irish Half Uncial) and results in a Structural joining of the letters. [Note: "Slanted Pen" Formal Hands are structurally separate]

THE STRAIGHT-PEN HANDS are generally written between ruled lines; this tends to flatten the tops and "pull" the curves of the letters, and contributes to their roundness and formality. Note: Appropriate and workmanlike RULING is of the first importance in Manuscript work.

CONSTRUCTION IN PEN STROKES

Note: most Strokes begin as Down-strokes.

"PULLED" CURVE in acdegoq &c. cao ce clqq carried further in C&/&n joins.

TRIANGULAR HEAD as stems of bdhijklu &c. (+) It is made with a quadrant (or + O) stroke.

UP STROKE used as stems of bhmnpr. bhp mnr ("This keeps the arches round")

POINT of NIB used in fine finishing strokes. fz Jg

NOTE. THE "SLANTED-PEN" gives better "dots" for ¡ and the "semicolon" + dot (¡) It is used also for dashes &c. The Pen may be very slightly slanted to give a better shape to v, w, x, y.

PLATE 4.—This MODERN HALF-UNCIAL has been used by me as a "copy-book" hand for students of Penmanship since 1899, being gradually modified to its present form. Its essential roundness & formality discipline the hand. Its elegance (due to the gradation in its horizontal curves) has an æsthetic value and fits it for certain MS. work, but unfits it for many *practical* uses where thin parts are liable to damage (e.g. as a model for type or letters formed in any material, or to be read at a distance. For such purposes the "slanted-pen" character is better; see Plate 6).—E. J.

It is in effect the "straight-pen" forms of the "roman" *small-letter* (that is, practically, the Roman Half-Uncial) *with the simplest necessary finishing-strokes added, and its general character assimilated to the Irish Half-Uncial of the Book of Kells.* It therefore represents the ancestral type of small-letter, and is a good basis for the later hands. (W. & L., p. 304.)

ABCDEFGHILM
MNOPQRSTVX
&JKUWYZ

The Capitals A to X are freely *written* copies of 10th. Century forms. They are here made rather lighter (with the exception of A.) & the lower thin Terminals in C E F L are turned down : otherwise they are practically identical with their originals. &c. to Z. are modern pen-made forms to match

ᚩ d e g h k m n s t u A G R S R

The Capitals O to U are freely *written* copies of 'round' forms in use in MSS. from the 10th. to the 12th. Century. (A to R are variants — based on early forms)

CONSTRUCTION. "Versal" Letters as here given are Pen-made, Built-Up, Roman Letters, *very freely written* generally in pure colour (*Red, Blue & Green*), or gilded.
Pen-made :— Written swiftly with a sharply cut, broad nib (*generally a little narrower than the Script pen — see footnote*) and left untouched.
Built-Up :— Each Letter is made with a given number of pen strokes — as sharp & clean as possible — and filled in : thus (A 3, B 3, C 7, E 10, &c):

In making horizontals in A & H (which are the full width of nib), & arms in C E F G L S T Z, the Pen-shaft is held horizontally.

For making the thick 'Bows' in B, C, D, &c the easiest & best general method is to make 1st the Inner curve (rather flat), & add the outer.

Roman Letters :—primarily 'Roman' (see Development): their slightly 'Gothic' character is acquired, & is due mainly to the use of the PEN; it may be increased by increasing the contrast of the THICKEST & THINNEST parts (a broader nib will do this naturally) and the curvature of the stems and forms. Or, on the other hand, their primarily Roman character may be brought out by diminishing the contrasts, & by making the forms more severe. This applies to both the 'SQUARE' & the 'ROUND' Characters. (Note: in the example of 'SQUARE' Capitals above I have emphasized their Roman Character by making the outer curves of B, C, D, G, O, P, Q, R, U, first, and adding the inner curves. E.J.)

DEVELOPMENT. Broadly, it may be said that the ROMAN CAPITAL FORMS were used for built-up Book-Initials & Headings till the 10th. Century ; the ROUND Writing-Forms (with the exception of the ornamental Capitals of the Irish School) being commonly subordinate — as VERSE-INITIALS — less formal, and more frankly written.
From the 10th. to the 12th. Century there was a departure from the severe Roman-form. It would seem that the Penmen — emphasizing the pen-character of the Initials — made greater use of the ROUND forms, & gave a general curvature to all the Letters (v. R). They also decorated them by flourishing tails or parts (v. b p a), and strengthened thin parts by the addition of decorative 'knobs' & buttress-strokes (v. b N A).

Form 12th. Century MSS.

After the 12th. Century the Initials were further curved and fattened and generally made more showy. Added ornament took the place of simple flourishing, and a type of "Illuminated Initial"— perhaps too well known — was the final outcome. Their treatment during this late development was often harmonious and beautiful, but they are too complex and too much compounded of their time for us to take as models.

PLATE 5.—"VERSAL" LETTERS, or early *Illuminator's Pen Capitals*, used in MSS. for Book-Initials (& Headings of Books), Chapter-Initials, and Paragraph or Verse-Initials, and even for coloured capitals in the text.
CONSTRUCTION : The width of the nib affects the character of the versal from the strongly *written* with a broad nib (*Gen. note*, Pl. 6) to the practically *drawn* with a narrow nib. Drawn or painted they properly acquire a different character. *Note : these examples show breaks and roughnesses that I had not skill to avoid, but left untouched as any trimming spoils their direct character.*—E.J.
The versal is the original of the Initial which has been the basis of Illumination for the last 1000 years. (W. & L., p. 205.)

I.

bcdfghilmnopqrux

(There is also in the original a hooked variety of m n & r)

(These 16 letters are practically identical with the forms of the originals, but are made LIGHTER in proportion: UPRIGHT (instead of slightly sloped): SHARP-HEADED (instead of blunt))

LETTERS COPIED FROM A 10th-CENTURY ENGLISH MS. { Harl. MS. 2904: Brit. Museum. } SLIGHTLY Modified

aest

forms suggested for model use instead of

aeſt

Original (Survivals) of Early Capital Characters xeſc)

jkvwyz

SIX additional letters made to match.

CONSTRUCTION OF "SLANTED-PEN" HANDS:
1. The position of the thick and thin strokes is approximately thus— It may be varied for different hands, but in any one hand should be nearly constant.
2. The STRONG oblique stroke should generally be emphasized, & dominate the curves, the WEAK oblique stroke should generally be suppressed and show only as a point.
3. The letters are practically un-coupled and their foot-hooks — as in the 10th. C. MS. are SMALL, HEAVY finishing-strokes (see feet of dhi &c.), except in t r where the hooks are essential parts (cf. l t, with their originals L r,).

DEVELOPMENT OF "SLANTED-PEN" HANDS:
This writing of the 10th. Century is derived from the early Roman (and Half-Uncial) forms modified by the French 9th. Century "Caroline" hands (to which it is closely related)
Et luminare minus ut præet noctæ estellas
EX. CAROLINE 9th.
and it is representative of the ancestral type from which have been developed THREE distinct and important TYPES (two of which may be regarded by us as permanent),

black letter (The early forms of this in Eng. & Ital. 12 C. MS are the best to study) and _italic_ and "roman" small letters

II.

abcdefghijklmnopqqrstuvwxyz

AN ITALIC HAND directly derived from the Foundational hand (I.) above.
The chief characteristics of the ITALIC HAND are 1. lateral compression, 2. branching of the parts (n u &c.). Secondary characteristics are 1. Elongated Stems, 2. a slight SLOPE. (This latter, probably the least essential, has been unduly exaggerated in modern use).

This example is made heavy to show the control of the pen (see General Note): various characters can be developed from it by (a) making lighter (b) making rounder (c) lengthening stems (d) flourishing e.g. bd hklp (semiformal 16th.) (e) coupling the letters (in less formal writing)

III.

abbcdefghijkllmnopq&

qrstuuvwxyz&123456789

(made with a "slanted-pen", for any of these hands)

A ROMAN-SMALL-LETTER HAND derived from the Foundational hand (I.) above, and assimilated to the Italian Formal 15–16 Century MSS. Various characters may be developed from this example by varying the weight & direction of PEN (e.g. the roman in plate 14, q.v., is a nearly straight-pen type).

GENERAL NOTE: A very broad nib strongly controls the letter & tends to give it a "gothic" character – due to the abrupt change from thick to thin. A narrow nib is more under the control of the Writer and its (gradually changing) stroke will give a more "roman" character to letters (the height being equal) nb/nb

PLATE 6.—"SLANTED-PEN" SMALL-LETTERS. Note: a "straight pen" form may be developed from these: cf. Plates 10, 14.
I. _Foundational Hand_: an excellent formal hand for MS. work and to develop into later forms (Ref. W. & L. collo. VIII. & pp. 305–310).
II. _Italic Hand_: a rapid and practical hand for modern MSS. (Ref. W. & L. collo. XXI. & pp. 311–315).
III. _Roman Small-Letter Hand_: suitable for the most formal modern MSS. (Ref. W. & L. collo. XX. & pp. 310, 481).
II. and III. may be taken as MS. models for practical adaptation to _printing, painting, carving_, &c.: cf. Pls. 10, 11, 14, 16.

School Copies and Examples, No. 2. John Hogg, 13 Paternoster Row, London.

Given the Essential-Forms or SKELETONS of the Roman CAPITALS, their finished character depends chiefly on the tool used to make them : ABCDEFGHIJKLMNOPQRSTUVWXYZ
it may further be varied by the selection of special forms (such as the 'Round' varieties - see 2.), and by various modes of finishing - with Terminal 'heads', 'feet', hooks, & flourishes (see 3.).

(1.) The simplest Terminal is a CROSS-STROKE as in these Skeletons ABCDEFGHIJKLMNOPQRSTUVWXYZ
[NOTE : the horizontals of B,D,E, &c take the place of cross-strokes likewise A,M,&,N,(V) may use their own strokes (or the left A,M,N)] Such letters may be called 'SQUARE' or 'CROSS-TOPPED': with some desirable inconsistencies, the Slanted-PEN forms are here given.

ABCDEFGHIJKLM
NOPQRSTUVWXYZ

(2.) ROUNDER forms (v. UNCIAL, pl.5) - finished generally with their own curves & hooks - have been developed from the Roman Capitals: these may be called the 'ROUND' or 'BRANCH-TOPPED' 'Letters: the Slanted-PEN forms are given here. [CE] BD & DEFGHR[U]MNPRSTUW

BDDEFGHKMNR
PSTU or for a more Formal MS. 'heads' & 'feet' may be added thus: BDDFHKNPRT [e.g. these would match very well with the letters of No. 1.]

(3.) FLOURISHED Letters are of every form and of every variety, simply drawn-out, or re-shaped, or added to: v. typical skeletons AAAA BB DCEFJGHJLMMNNPQRJUVELT & I-PEN forms below.

ABCDEFGJLMNRPELT
AAAB&JFHBMOTUVx
MLEF2XXX {EMQRVW}
Additional Varieties of Form not included in above Examples

General Note. These three "Alphabets" — SQUARE, ROUND, & FLOURISHED — are to be regarded as varying forms, to be used freely together of ONE Alphabet most prolific for the Penman. Provided the Pen-nib & the movements of it are kept uniform, the different forms become harmonious and they occur in the same MS. and EVEN in the same word.
The forms given above may be varied in every detail, & the alteration of their proportions (see Pl.9) or weight of pen-stroke (see Gen'l Note, Pl.6.) will further give untold varieties.

PLATE 7.—"SLANTED-PEN" CAPITALS, showing admirably the constructive power of the pen in making characteristic letters out of skeleton ABCs : they are not directly copied from or founded upon any given MS.
Note : In mediaeval times Capitals were not so much a necessary complement of small-letters, as a different and more important type of letter, used chiefly for large Initials and Headings : and a standard type of simple text capital to match a standard small-hand does not seem to have been recognised before the 15th century.
Any of these may be used (modified appropriately) with any slanted-pen small-letter, such as those in Pl. 6, or for MS. in capitals. As models—particularly in the method of their making—they will be suggestive for craftsmen generally.

FROM THAT TIME EVER SINCE,
the sad friends of Truth, such as durst
appear, imitating the carefull search
that Isis made for the mangl'd body
of Osiris, went up & down gathering
up limb by limb still as they could find
them. We have not yet found them all,
Lords and Commons, nor ever shall
doe, till her Masters second comming;
he shall bring together every joynt &
member, & shall mould them into an
immortall feature of lovelines &
perfection

PLATE 8.—ITALIC MANUSCRIPT, modern, from Pl. 6 (the Roman small-letter from Pl. 6, the Capitals from Pl. 7).

School Copies and Examples, No. 2. John Hogg, 13 Paternoster Row, London.

AABBCD DEFGHIJJ KLMNOP QRQSSTU VWXXYZ

PLATE 9.—PEN-MADE ROMAN CAPITALS. *Construction:* freely *written* (Nib-width = Thin-stem-width, *cf.* N). built up as versals are (see Pl. 5), but the serifs here *blend* with the stems, of which they are an actual (not an *added*) part. PROPORTIONS (CLASSICAL) : WIDE (width=abt. height) OCGDQ & HAMNT(U)VZ(W), NARROW (width=abt. ⅔ height) BEF(K)PRSXY & I(J)L : all may be varied. The Curves all follow O, which is here " upright." For occasional forms *any* parts may be drawn out (cf. *skeletons* (3) Pl. 7).
DEVELOPMENT : *The pen originally helped to characterise Roman Capitals by determining* " *thicks* " *&* " *thins* " : for further pen developments see Pls. 7 & 10.

School Copies and Examples, No. 2. John Hogg, 13 Paternoster Row, London.

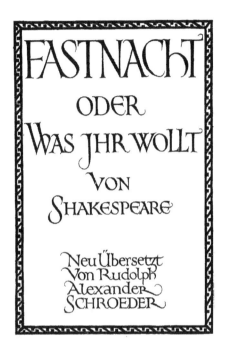

Plate 10.—A TITLE PAGE (actual size) (*by permission of Count Kessler*) drawn with a pen by E. Johnston and engraved in wood by Noel Rooke. The Capitals are the pen forms of Plate 9, drawn carefully, with ornamental developments suggested by the pen. The small letters are an *upright* form, built-up and developed ornamentally, from slanted-pen hand III., Plate 6.

Title pages and similar subjects should either be engraved in wood or metal or be printed in good type (see Plate 11). As constituting a decorative part of a book, a " processed design " is generally unsatisfactory.

PLATE 14.—"LOWER-CASE," ITALICS & NUMERALS incised with a "V" section (see Plate, of Capitals, 13) in Hopton Wood Stone, by A. E. R. Gill. These letters are appropriate for all ordinary inscriptions in stone. While they are as easily and quickly made as the more common "sans-serif" or "block" letters, they are at the same time more legible. (See Plate 6 here, and W. & L., p. 310, etc.)

INCHES

PLATE 15.—"RAISED" LETTERS—CAPITALS & NUMERALS—carved in Hopton Wood Stone, by A. E. R. Gill. The section used, ⟍⟋, is the easiest to cut, the strongest and generally the best for ordinary work. The forms and sections of the large letters at the foot, ▬ & ◢, are only appropriate for isolated letters or words or for obviously ornamental uses. Raised letters in stone are more of the nature of "carving" than "writing," and are less suitable for ordinary inscriptions. (See W. & L., pp. 377, 403.)

The Artistic Crafts Series of Technical Handbooks.

Edited by W. R. LETHABY.

"We would have this Series put Artistic Craftsmanship before people as furnishing reasonable occupation for those who would gain a livelihood. . . . In the blending of handwork and thought in such arts as we propose to deal with, happy careers may be found as far removed from the dreary routine of hack labour as from the terrible uncertainty of academic art. It is desirable in every way that men of good education should be brought back into the productive crafts : there are more than enough of us 'in the City,' and it is probable that more consideration will be given in this century than in the last to Design and Workmanship."—*Vide* EDITOR'S PREFACE.

BOOKBINDING, AND THE CARE OF BOOKS. A TEXT-BOOK FOR BOOKBINDERS AND LIBRARIANS. BY DOUGLAS COCKERELL. With 122 Diagrams and Illustrations by Noel Rooke, and 8 Pages of Collotype Reproductions. 352 Pages. Price 5s. net. (*Second Edition.*)

"An excellent book opens 'the Artistic Crafts Series of Technical Handbooks.'"—*Times.*

"It would be hard to find any technical book of this kind which gives more importance to considerations of good taste than this. . . . It leaves no part of its subject unaccounted for, and is in its own printing, binding, and illustration, a favourable example of good craftsmanship. Valuable in itself, it promises well for the series which it opens."—*Scotsman.*

SILVERWORK AND JEWELLERY. A TEXT-BOOK FOR STUDENTS AND WORKERS IN METAL. BY H. WILSON. With 182 Diagrams by the Author, and 16 Pages of Collotype Reproductions. 348 Pages. Price 5s. net.

"A lucid text-book for students and workers, well illustrated, being the second volume in the series which made a successful opening with Mr. D. Cockerell's 'Bookbinding.'"—*Times.*

"It teaches not only processes and workshop practice, but also good taste in the making of objects in which in these days vulgarity is none too seldom seen. Admirably illustrated, well written, and practically serviceable, the book should prove welcome alike to craftsmen and to amateurs."—*Scotsman.*

". . . Will have a fascination for all craftsmen, and may be read with advantage by every one who wishes to understand the underlying principles of the art crafts."—*Morning Post.*

"We cannot imagine a better aid and supplement to practical experience in the workshop than this handbook. All necessary processes, from the simplest to the most complex, are explained in a methodical and logical order, with the aid of illustrations, in which each touch is eloquent and fresh from the master's hand."—*Athenæum.*

STAINED GLASS WORK. A TEXT-BOOK FOR STUDENTS AND WORKERS IN GLASS. BY C. W. WHALL. With 72 Diagrams by two of his Apprentices, and 16 Pages of Collotype Reproductions. 392 Pages. Price 5s. net.

"Mr. Whall addresses four classes : the workers, the artist-amateur without any technical skill, the patron, and the church architect. . . . A very careful, concise, and artistic work."—*Pall Mall Gazette.*

"Fascinating and instructive. . . . The book should also prove of great service to those interested in painted glass, whether as custodians of buildings containing fine old windows or as architects or clients intent on securing good modern work."—*Glasgow Herald.*

"The highest praise possible for this handbook would be to say that it is worthy of the remarkable series to which it belongs, and this, without reserve, we can affirm to be the case. Mr. Whall is a veteran at his craft."—*Arts and Crafts.*

WOODCARVING : DESIGN AND WORKMANSHIP. BY GEORGE JACK. With 78 Drawings by the Author, and 16 Pages of Collotype Reproductions. 320 Pages. Price 5s. net.

"Those who practise wood-carving will find in this admirably written and illustrated book a comprehensive treatise on the subject."—*Morning Post.*

"The illustrations complete in a very appreciable way the value of one of the soundest text-books within the reach of the student of woodcarving."—*Glasgow Herald.*

"His illustrations from both ancient examples and from his own work are excellently chosen and reproduced, and show that he has made himself the master and the pupil of the best traditions of his craft."—*Pall Mall Gazette.*

EMBROIDERY AND TAPESTRY WEAVING. A PRACTICAL TEXT-BOOK OF DESIGN AND WORKMANSHIP. BY MRS. ARCHIBALD H. CHRISTIE. With 187 Illustrations and Diagrams by the Author, and 16 Pages of Collotype Reproductions. 416 Pages. Price 6s. net.

"There are various publications already in existence from which one may learn something of the mysteries of fine embroidery, but we know nothing so wide in its scope and full in necessary detail as Mrs. Christie's."—*Birmingham Daily Post.*

"The illustrations are certainly designed to simplify to the student the details of the work in the clearest and most simple form, and the letterpress is equally explicit. So successful is the author in this respect, that any clever needlewoman might easily, without any other aid, acquire executive skill in this ancient and beautiful art."—*Auckland Star.*

WRITING & ILLUMINATING, AND LETTERING. BY E. JOHNSTON. With 227 Diagrams and Illustrations by the Author and Noel Rooke, 8 Pages of Examples in Red and Black, and 24 Pages of Collotype Reproductions. 512 Pages. Price 6s. 6d. net. (*Second Edition.*)

"Is itself a striking example of artistic craft, containing a wealth of beautiful illustrations, and being produced altogether in a style that makes it a pleasure to handle the book. Mr. Johnston has dealt with his subjects in the most complete manner, so much so that any one who thoroughly masters this volume will know pretty well all that there is to know about Writing, Illuminating, and Lettering."—*Newcastle Journal.*

"It would be almost impossible to read this artistically conceived and executed volume without gaining benefit beyond anticipation."—*Sheffield Independent.*

"No mere writing will fittingly describe this delightful book."—*Art Teachers' Guild Record.*

"It may be doubted if any one has ever before so carefully explained how to write."—*Tribune.*

FURTHER VOLUMES IN ACTIVE PREPARATION.

SCHOOL COPIES AND EXAMPLES. SELECTED BY W. R. LETHABY AND A. H. CHRISTIE. 12 Drawing-copies (1 in colours), 15¾ × 12, with Descriptive Letterpress, in a Portfolio. Price 5s. net.

*** Projected as a Standard Series of Handbooks on the Artistic Crafts, suitable for Schools, Workshops, Libraries, and all interested in the Arts. Each Craft will be dealt with by an Expert qualified to speak with authority on Design as well as on Workmanship.

Published by JOHN HOGG, 13 Paternoster Row, London, E.C.

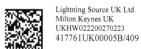

Lightning Source UK Ltd.
Milton Keynes UK
UKHW022200270223
417761UK00005B/409